MY STORY

by

Carole Anne Murphy

To my husband David, who stood by me through all the hard times.
"I love you ."

My story begins in a small town outside of Sydney, Nova Scotia where I was born and lived for part of his life. I had wonderful parents, my Dad worked for the Canadian National Railways and my Mom stayed at home and took care of us, my brother Roger, who was the oldest and my sister Sharon who is the youngest. She was a wonderful Mother, always there for us, assisting us with our homework, teaching us right from wrong, cooking wonderful meals, and dressing us with the latest fashions. I never once supposed, I would ever leave home. I wanted always to be with my Mother and Father. I loved them dearly. I assumed they would always be there. The sadness I would feel when they passed away was devastating still to this day. They are never far from my thoughts. I was always a very emotional child and always seemed to be sad.

I suffered from having bad nightmares a lot. I would leave my home and roam around the region in my Nightclothes, summer or winter, it didn't matter. I would wake up when the stones hurt my feet or the snow was too cold, then I would go home and then go back to bed. I don't believe my Mother knew every time I was out. To this day I still have nightmares and wake up hollering for my Mother. I am afraid to go to sleep because I know that soon I will have another episode.

I graduated in high school in 1958 and started my career as a secretary working for a chain of jewelry stores. I was gainfully employed there for three years.

During this time I met my future husband. Jake was every bit as good looking as my sister said he was. He was blonde and blue eyed with a muscular build.It wasn't long before I fell in love with him and strongly thought only of spending my life with him. He was every young girl's dream. We dated for about a year when he decided to join the Royal Canadian Air Force. There wasn't a lot of work in Sydney, other than the steel plant and the mines. His father worked at the civilian plant for over 45 years, and he really didn't want Jake to do that for the rest of his life.

It wasn't long after he enlisted that he had to go to St. Jeans, Quebec and Aylmer, Ontario for his basic training. After his training, he was posted to Comox, British Columbia. He didn't come home on any leave for well over a year. It seemed forever since I saw him, and I missed, him oh so much. I wrote to him everyday. My life then was work and stay home and wait for him. On Sundays I would go to his parents home for the day. I felt better there because I knew how much his mother missed him, and we had this in common. For whatever reasons, we barely spoke on the phone, whether it was too expensive or what, I really didn't know. The one phone call I did get was when he asked me to marry him, and of course I said yes. I was so happy I wanted to spend the rest of my life with him. I loved him with my whole heart, and I dreamed about having boys who would be just like him. I wanted to have a baby, as soon as we could. He wanted me to pick a date for the following date for the following summer, and then he would arrange for his leave. I picked July 5, 1961 for my special day. He sent me my engagement ring through the mail. I was so proud to wear it. I did have so much to look forward to, also I was beginning to genuinely think about my having to leave my devoted family and move so far away. It was from the East Coast and the West Coast and to me it seemed like the other end of the world.

I had never been away from Cape Breton in my 21 years, and it was scary. I was all torn up, but I knew that was what I wanted. I wanted to be married and have children. I could go home for the holidays so it wouldn't be so bad.

My Mother said that I was born in a veil. That was a sign of having luck in my life. If only that had been true.

My future Mother- in –law was a seamstress, and she offered to make my dress and also the attendants dresses. We looked at wedding books, and I picked out what a liked. Jake's parents were polish, and I knew they wanted a big wedding. I was to have five women and five men in the wedding party. The colors were picked out and what the men would be wearing. I didn't have much to say in it, or it seemed to be that way. There were lots of planning for my big wedding. My dress was being made early because there was a lot of work to be done for the other five dresses. I had many fittings and finally my dress was made. It was beautiful, and I felt like a princess when I put it on. I could hardly wait.

The day finally arrived, and I was feeling sad. I didn't know why, but I was. Our house was full of out of town guests. Everyone was in such a happy mood. They were mostly my Mom and Dad's family, and they loved to party. Everyone wanted to take pictures, so I spent a few hours out on our lawn. It was so hot that day I worried about my hair coming undone. My Dad had said it was too late to change my mind. I don't think he wanted his little girl to leave home.

We were married in a small polish church and so many people came. I was nervous but happy. We had a big reception, lots of polish food and liquor. We heard the party went on for a week while we were on our honeymoon.

We stayed in a log cabin motel which I saw being built when I was a young girl and wanted someday to spend my first night there so we did.

We had borrowed my Dad's car and drove to Halifax. We couldn't get an accessible room anywhere so we ended up staying at my Grandmothers for the night. My husband slept on a sofa downstairs and I was upstairs, how romantic.

We drove back to Cape Breton, so we could spend a few days over Cabot Trail. We found a motel and that night we finally did make love. It was the first time, since I met him that we made love. Was this it I assumed? He decided the next day to go home and spend time with our families before we moved to British Columbia. I wished then that I had stayed at home.

He dropped me and the car off at my Mom and Dads and eventually took a taxi to his place. I guess the honeymoon was over. He stayed the next week with his parents, and I stayed with mine. My God I assumed is this what marriage is all about. I was happier with my life being single. The day came when we had to go to the airport to catch our plane to the West Coast. I cried so much leaving my Mom and Dad. I didn't know where we were going, what it would be like there and leaving my friends. It was all over powering to me, and I was afraid.

We landed in Vancouver, got a hotel room and left shortly after to go to dinner and take a walk through downtown Vancouver to see the impressive sights. We ended up at an all night movie house. The next morning we were to catch a ferry to Nanimo where friends of Jake's were picking us up and would then drive us to Comox. The conversation was all about sports and how he couldn't wait to get back and start again. I didn't know he was so involved in sports. There was a lot I didn't know, but would soon find out.

When we arrived at our little house that Jake rented us, everybody came in. when some friends of his came in, they carried on like it was their secluded house. I wondered what was going on. I didn't get a chance to ask Jake if it was our place or what. For supper we had lunch which the woman prepared. I don't even remember what her name was. Everyone got up and got ready to go to the base track and a field meet. I was asked to mind their baby while they went. I finally got Jake in the nearby bathroom and asked him what was going on, and if this was our house or what. He said it was, but they would be staying until the movers came in to move their furniture. I was afraid to ask when that would be, so I didn't ask.

Around 10:30 that night they came home, watched t.v. for some time and then went to bed. My husband and myself were to sleep in the living room on a pullout chesterfield. You could hear the people in the bedroom breathe and that was with the door closed. I missed my family, and I wanted to go back home where I was loved and cared about.

The only way to get off this island was by ferry or to fly. I had no money for myself, so I was stuck. I had no telephone. What would that matter, I wouldn't tell my parents that this was a big mistake? I loved Jake and things would be better.

Woody and his wife had their furniture picked up so Jake, and I went to Courtney to get a bank loan, so we could buy our furniture. I was so happy and excited. Finally, I would have my husband for my own, more or less I thought.

The next day my furniture arrived. I was arranging everything while Jake left to play a ball game. Him being a sports junky that he was, he would dominate my whole life. I heard a commotion in the kitchen and went out to see what was happening. My guests were back again; ready to spend the night using my new bed that I haven't even slept in yet. Again, I slept in the living room. I was anxious for my husband to come home. I had things to say to him. Get rid of them or I was leaving. If I had to walk it didn't matter, I wanted out.

The next morning I was eating "my" breakfast, I saw suitcases and boxes being loaded into their car. They were leaving, hurrah. They left without not even a so long . One month guests, I was finally alone at last.

I wasn't feeling too good for a few weeks, symptoms like the flu. After six weeks I strongly thought that I should find a doctor and go to see him. I had my checkup a week later. I had to go for a follow up new appointment. During that visit I told the doctor I still had the flu he said no, your going to have a baby. The second time I had sex with my husband must have got me pregnant. I was happy but Jake didn't have much to say. Cannot remember if there was a celebration, but it's doubtful. I wanted a healthy boy, and I couldn't wait . I needed someone in my life. Someone to keep me company and fill the lonely hours when a basketball game had a preference over me.

It was a beautiful summer day, and I had finished all my housework, so I decided to go outside and lay in the sun. it wasn't very long when a neighbor came over and introduced himself. He told me it wasn't a good idea to lay in the grass because there were very big spiders around. Being terrified of spiders I ran into the my home, never to lay in the sun again.

I met a married couple who Jake knew. Bob and Millie were their names. I liked them, as soon as I met them. We visited back and fourth. One evening while at their PMQ, (Private Marriage Quarter's) Bob told me to come into the laundry room, and that he had something to show me. I saw the biggest spider I have ever seen in my life. One evening Jake home with a little Collie pup for me. He was adorable, and I called him Shep. I put newspapers down in the kitchen, so I could paper train him. A few minutes later he started to bark. I got up to see if he was okay, when I saw the paper move across the kitchen floor. There was the biggest spider crawling across the floor. I jumped from the laminate floor to the top of the kitchen table and stayed there until Jake came home and got rid of it. After I jumped, I swore the baby's feet went under my ribs and stayed there until he was born. The pain was there until his birth.

A few months later a posting came for Halifax, Nova Scotia. I was happy, it would bring me closer to my family. I wrote my Mom right away. I knew she would be happy. The posting came in at a good time because in a few months I wouldn't be able to fly on account of the baby.

We arrived at Halifax and took the different train to Sydney. I was going home and couldn't wait. My Dad picked us up at the station. Again Jake went to stay with his parents. That was okay with me. Mom asked where he was, and I told her that he went home. She offered him a decent place to stay, but he said it was okay. I stayed with my Mom while he went to Halifax to look for a home for us . A few weeks later he got us a small one- bedroom apartment in a rundown house. When our furniture arrived, we moved in together. My Mom and

Dad came up because I was near my due date, and they wanted to be with me. It was a little cramped, but we managed. They were always there for me. They seemed excited about the baby. It would be their second grandchild. They went shopping and arrived home with a crib and carriage. I could see my little baby in the crib. I wanted to have him right away, and it was hard to wait.

After two weeks and my Mom and Dad patiently waiting for the baby, it started to happen. I went to bed early feeling tired. Around 11:00 that night my water broke, and I was on my way to the hospital. There were a lot of expectant mothers lined up in the hallway. I had a lot of pain but only in my back. Jake was there with me. I knew he should go home, so he could get up for work in the morning. He strongly thought that was a good idea, so he left. At five to seven in the morning our son was born. The little guy would be so wanted and loved. I wanted to be a good mother and looking back over the past 40 years, I couldn't have been any better. He was my whole world and nothing else mattered.

After I returned to my room, I slept for a few hours, waking up when I heard the babies crying as they were given to their mothers. I didn't even know that he had been born on Mother's Day, May13, 1962. I wanted to hold him and check him all over, ten fingers and ten toes. I sat up in bed waiting but no baby came. I rang for the nurse and asked her where my son was. She told me that he would be down later, and that they were monitoring him, for what I didn't know.

I called Jake and asked him to come down and see why our son was not brought down from the nursery. It was hours before he came in, or it seemed like it. They told him the baby wouldn't be down until the next day. I got in a wheelchair and made him take me up to the nursery. I wanted to see our son. I peered through the window and could see him in the distance. I beckoned the nurse to bring him over. He was crying so much, and I just wanted to hold and comfort him. He was all bundled up in a blanket and looked so tiny, but I fell in love with him right away and just knew that my life would now be better.

Nobody said anything to me the next morning. They brought him in, and I held him in my arms, never wanting to let him go. Jake was there, kind of quiet not seeming to be excited when the doctor came in to talk to us. I would soon find out why. He removed the blanket from him and my eyes immediately went to his twisted little feet. I couldn't believe what I was seeing. The doctor whose name was Dr. Sinclair was an orthopedic surgeon, he said that Paul had clubbed feet and that he would be taking care of him. He would need to put casts on right away. I was in shock, and I cried and cried until I couldn't cry anymore. I put all my faith and trust into this doctor. The next day my baby came to me with tiny little casts on his feet. I asked God why did this happen to my baby. It just wasn't fair.

After a week of being in the hospital, it was time to go home. My Mom and Dad secretly saw his little casts through tears in their eyes. It was sad for them.

I named my son Joseph Paul. He would be called Paul after Jake's father. I liked him a lot and since it was his first grandchild, I genuinely thought it would be nice to call our son after him. He seemed proud and happy when we told him. His grandmother said " no way was she calling him Paul." I still named him Paul with or without her approval.

Paul cried a lot. I would hold him close to me to comfort him. Maybe he's casts were bothering him. I phoned the physician to make an appointment. He had a clinic in the hospital, so I would be seen there. The doctor explained what he would be doing to straighten Paul's feet. He told me he would need surgery when he was older. The casts would duly stay on for 6 months and then he would go into braces.

I had to have him checked every week. It meant spending hours at the clinic, and he cried his little heart out. I didn't know why he cried so much. They would eventually take him in the office to apply the cast while I was in the waiting room, and I could hear him screaming. I walked the floors waiting for him. A lady told me he would be okay. She went through it every week. I didn't want them to hurt my baby. They eventually brought him to me. I held him so tight to my heart. I didn't know how we were going to go to through this every week, but we would, and we did. His little feet were so deformed; anything they could do to straighten them would be alright with me. It would be a long road, but I was always willing to do anything.

After six months he would be put in orthopedic shoes with a bar attached to the bottom of his shoes. We would have to do exercises with him twice daily, then put the shoes on him. He had to sleep with shoes and the bar on. It was so uncomfortable looking. He couldn't roll over but he sure could hit the bars of his crib, cracking them. All night long he would be banging the crib. I would eventually pick him up and hold him.

This went on all night long, and I was getting tired with no sleep. Paul did nothing but cry. I didn't know why, and it was like, he was in pain. The doctor said it shouldn't be that uncomfortable for him and that he would get used to it. I wanted to take the shoes and bar and throw them in the garbage. It was so unfair. I cried as much as he did.

My Mom and Dad had gone home. I sure missed their help. It seemed that I was the only one doing anything for our son. I realize that Jake had to work during the day but what about at night. Instead of playing basketball and baseball in the evenings, why not help me? Even so, that was not to be.

One afternoon a knock came at my door. When I opened the door there was his sister, bag and baggage. I asked her where she was going, and she said didn't Jake tell you that I would be staying here. No I said, Jake didn't tell me anything like that. I wonder why he forgot. Needless to say when he came home, I found out. Seems that she was pregnant and had to leave home because her father wasn't to know, and she would be staying with us until the baby was born. It was hard enough in a one- bedroom apartment basement apartment, with a baby that never stopped crying. What could I do? She was staying and that was it. She slept in the tiny living room with the chesterfield pulled out. There was no room to walk around it. All night Paul cried and kept everyone awake, but at least she got to sleep through half the next day while I took him for a scheduled walk. She never did a thing to help me. I had Paul to take care of, laundry to do, meals to make, and I was wearing out. I wondered why I was so burdened. I didn't want this kind of life. It was hard enough to take care of my baby let alone other distractions.

My head never stopped pounding, and I never stopped crying. I was only 21 years old, and I should have been out having fun. All my energy went into my baby. He didn't ask to be born. It seemed I was the only one who cared. I never had a minute to myself.

I made an immediate appointment to see my doctor. I told him that Paul never stopped crying. He put him on gripe water genuinely thinking he had colic. Unfortunately, he didn't have colic so the gripe water didn't help.

In the meantime I still had my guest. She knew I didn't want her there so things were pretty tense. Her father used to call, and I was to tell him that Carole was working instead of having a baby.

She was getting big and did nothing but sleep. How I wished I could. One day I woke up and said enough was enough. You are not staying another day. I called a home for unwed mothers and got her a accessible room. She was to be there right away. When Jake came home, he couldn't believe that I had done this. How could I be so uncaring? He hugged and held her how I wished I had someone to hug me and comfort me. Because our bedroom didn't have a door and there was no privacy, we couldn't have sex. This must have made him happy. I don't know why he was putting up obstacles. Such a handsome young man, what was his problem . Was I just hired to be his maid, with no fringe benefits ! I'm sure there was some kind if problem, because over the 20 years we were together nothing much changed.

Paul was having his appointments and doing better. I could see a little change in his feet. All the hard work was starting to gradually help my little boy. He went through so much. I loved Paul with all my heart and didn't care what I had to go through to help him. He was a gorgeous looking baby; blonde hair and blue eyes, just like his daddy.

One day Jake called and said he could have a two-bedroom apartment in Dartmouth, just across the bridge from Halifax. I was so happy we would be moving. I had enough of living in a damp, cramped basement apartment. I must have had blinders on when I moved into this place. We got settled into our new apartment. Paul had his own room. I decorated it for him.

One day he seemed to be quiet, so I went to his room to check on him. He was lying on his back. I had taken off his bar and was going to get him up and exercise his feet. I noticed that he was drawing his legs up to his chest and appeared to be holding his breath. I got him up and everything seemed to be okay. A neighbor had dropped in with her nine-month-old baby, and I noticed that Paul could not sit up like her baby. I had brought it up so many times to the doctor, and I was told that he was a big lazy baby and the shoes and the bar were not helping him.

That night after bathing him I laid him on a blanket in the living room on his stomach. I noticed that he was having trouble holding his head up and his head would drop, and he would hit his forehead on the laminate floor. He cried his little heart out, and I did too. I held him for a longtime and laid him on my lap. As he was lying there, he drew up his legs, and it seemed like his eyes were rolling back in his head. I was scared. What was happening to my baby? He was 9 months old and couldn't do anything but cry. When I took him to the clinic I asked the doctor if the brace would stop him from sitting up, and he said no. A lot of his babies with a bar managed quite well and could ever crawl with it. My Mom and Dad came up for the weekend. I was so happy to see them. She said I had lost weight and look tired. She knew I wasn't getting much sleep.

That evening after she bathed Paul, I took him and laid him on a blanket. Again I noticed that he was drawing his legs and was losing control over his eyes. I mentioned this to Dad to watch him, and he saw it and said to get him to the doctor, that he was having a seizure. I laid him on his stomach, and right away he dropped his head and banged his forehead. My Mom assumed that he was just tired and dropped his head. I wondered why this was happening. My darling little boy, what else could go wrong? Hadn't he had enough in his short little life?

I made another appointment to see the doctor. Paul would never drop his head or roll his eyes in the presence of a doctor. He only did it when we were alone. I told the doctor that I didn't imagine it. He was doing it more and more everyday. I wanted to know why at 10 months he couldn't sit or roll over. Again I was told that he was a lazy little boy. I wanted to see another doctor but according to him it wasn't necessary.

I took him home and tired to cope with this. It wasn't easy. It was time to buy another crib as he had broken all the rungs on his. We had to nail his crib to the floor, so he couldn't move it all over the floor with his banging. His little forehead was always bruised from banging it so much. Every time I heard him; I would hold him and the bad thing passed. When I would look into his eyes, it was like; there was nothing there.

He wasn't a happy baby. He made no sounds, or attempted to say Mamma or Dadda, but he was a beautiful baby, and I loved him with all my heart. I wanted him to crawl and walk and for this terrible thing to go away.My head ached all the time, and I was so tired and depressed. I had no friends;I didn't have time for anything.

I kept on with his foot exercises and his appointments to have he's casts put on. I could hopefully take the shoes off for longer periods of time, but at nighttime; he had to have them on. He seemed to be getting used to them. He still banged the rungs on his next crib. I wondered how long this one would last.I was giving him a bath when he started pulling his legs up and this time I saw froth coming from his mouth. His eyes rolled all around, and it was like; he was holding his breath. I got him dressed, and we left for the doctor's office. I didn't care; someone was going to do something if I had to sit in the office all day. I was staying until my son did this and the doctor saw it. It never lasted too long but a few minutes were a lifetime to me. The receptionist said his doctor wasn't in but the doctor on call would see him. His name was Dr. Mc Kinnon. I felt I was going to get something done this instant. We waited for quite a while. The doctor was young, but he was interested in what was happening. Paul didn't have another episode but the doctor positively thought he should be admitted to the hospital for observation. He was admitted that day. I was so relieved but so scared. I missed my little boy, but I kept busy during this time away. I didn't sleep much. I was wondering how he was, if he was crying and had no one to hold him. All the terrible thoughts I had. Jake was worried like I was but had his job and that occupied his time. I talked to him a lot during this time, and he was hurting; I could tell.

It was early afternoon when the phone rang with a call from the doctor. I was shaking so bad that I needed some good news but there wasn't to be any. The particular tests were completed, and I was told that Paul had Phenylketonuria (PKU). I couldn't understand what he was really saying until he told me that Paul had a lot of brain damage done because of this disease. My God, I was thrown into a black pit and didn't know if I would ever get out. The pain and sorrow I felt for my son were devastating. I cried and cried and walked the floors. I couldn't believe that Paul was mentally retarded. Those were the words that described my baby at that time. What did I ever do to deserve this? I wanted a normal little boy. I don't remember much after that. My husband came home after he got his call. We were both clung to each other. He was as broken up as I was.

An appointment was made to go and see the doctor where we could find out about this terrible disease. PKU is a genetic defect involving the metabolism of an amino acid, in this case phenylalanine. The primary defect is the inability to convert excess phenylalanie from food resulting in an accumulation of the phenylanine in the blood. On account of the regular baby food or milk that Paul was given, it was doing damage to his brain which in most cases causes severe retardation. Seizures may occur, tremors and hyper activity. PKU is most common among Northern Europeans and Italians.

Since the early 1960's screening newborns for this disease have been widespread. Today all hospitals in the province are required by law to test newborns.

PKU is a rare disease, which can be controlled by a strict diet. The diet must begin in the first few weeks of life for normal brain development. Because Paul was 14 months old he had a lot of brain damage done. Since all foods contain phenylalanie , none can be selectively chosen for a low pheylalanine diet. With therapy, mental development and body growth are normal and the other side effects of PKU are eased.

When I left the hospital with Paul when he was born, I was suppose to have been given a diaper disk that would have shown that Paul had PKU. This was the test that was suppose to be done before they started testing with blood from the heel of the baby. I don't recall seeing a disk that I was suppose to put in his diaper and mail it back to the hospital. I would have to live with this for the rest of my life. A disk would have detected PKU in Paul.

They kept Paul in the hospital for awhile longer until they got him on his diet. The milk that he would have to drink came from the United States. It was powdered form. It cost $100.00 for 12 cans which would last a month. That was a lot of money in those days. It passed as a drug; therefore, we didn't have to pay for it. People who didn't have a medical plan could not afford the milk. It was terrible tasting but Paul drank it with no problem. It was important that he drank so many ounces but no more than what was prescribed.

The day came when it was time to take him home. We had to see a dietician, learn about he diet and how of the highest importance it was and not be given anything that wasn't on the diet. He would have to have blood work done after to make sure his level of phenylalanine was in the normal range. If it was low then I was doing my job. It was an awful burden, and it weighed so heavy on my heart. I held him so tight when I saw my baby. I missed him, and it was wonderful to have him back I told him he would get better and that I would see to that.

Paul was on his diet for about 3 months when I noticed that when I laid him on his stomach, he had control over his head and the seizures seemed to be less. He still cried a lot and was up a lot at night. One day I went into his room as I heard a terrible banging. He was standing up banging his forehead on his crib. I couldn't believe that he could stand; it was a miracle. How a baby who couldn't sit, all of a sudden could stand holding onto the crib. His forehead was red and swollen. Why he banged his head so much I didn't know .The diet he was on consisted of 2 table spoons of pears or 1 tablespoon of squash and a bit of pabulum. This was his food for the day, along with his formula. He seemed to be hungry, but I couldn't give him any more even though I wanted to. The formula would serve to fill him; I hoped so. There were a lot of appointments and blood work. He cried so much when we went to the clinic. When they took him from me, I could hear his crying and terrible screaming. Many a time I walked into the office to see what they were doing to him only to be told to leave. I dreaded those visits, but it had to be done. Sometimes the levels got out of range, so his formula had to be lowered. It was a never-ending battle, but it was for the betterment of his life.

In 1964 I found out that I was pregnant. I had such mixed emotions. What if this baby had the awful disease, but also what if he did not?

I was handling Paul okay. His appointments kept me busy. One day he sat up by himself. It was a joyous time. He still banged his head and would shake his crib so much; he could move it all over his room.

One day he managed to get it over to the window and banged his head on the glass and broke it. He was lucky he didn't cut his face. Jake had to get a frame made to enclose the window. He went back many times to pull on the frame trying to move it.

He started to climb out of the crib, half the time falling on the floor with a crash. He would take his bedclothes off in bed, and remove the vent that was in the floor and put all his bed linen in the vent. It was like overnight he started doing all these things. Just like a little monster.

He never stopped banging his head. I put a helmet on his, and he would take it off. Another crib was about to go out to the garbage. I didn't accept the part of his brain being damaged. I guess I just blocked it out.

As time went by, I was looking forward to my baby coming. I so wanted a little boy without PKU. I had such dreams. On October 22, 1965, I had another little boy, I didn't see him that day. The cord was wrapped around his neck, and he had little star marks on his face. The doctor came and said the next day it would be gone and I could see him then. He assured me of that. I asked if he had PKU and all he said was, " they would do the test later that day" unfortunately , he had PKU. I swallowed hard and said okay, it was picked up at birth; I knew I could take care of him. I would make sure he stayed on his diet and everything would be okay. I knew about the diet, and what he could have and not have. It would be difficult, and I had confidence in myself. My boys for 6 years would never eat any kind of dairy products, no sweets, no meat, fish, or bread. Nothing but a few teaspoons of fruit and a bit of squash, along with their powdered milk.

I was so lucky that Paul and Kevin liked their milk. If they finished early, then they had to go without it until the next day. Mothers would tell me they had to spoon feed their children because they didn't like it. We couldn't eat our own meals in front of Paul. It was so unfair, and I knew he cried a lot because he was hungry.

As Paul got older, the cupboards would have to be tied up, also the fridge door. I couldn't chance him getting any food.

My Mom and Dad came up to see Kevin and Paul. It was always nice when they came to visit. She was so helpful. If only she had lived a little closer to me. I had no one from my family around me, and It was so hard to cope sometimes. The left side of my head ached a lot, and I hoped I didn't have a brain tumor. It was strange why the pain was there so much. I didn't go to the doctor as I was afraid of them. They always seemed to give me bad news.

The doctors that Kevin and Paul were seeing suggested that we try to get to Toronto to the sick kids hospital. They were so much more advanced there, and it would be good for the boys. Jake looked into getting posted there. A month had gone by and it hadn't heard anything, so he inquired again. Eventually, he got one, but it was in Hamilton, Ontario. He took it. I was sad because it meant moving away from my family, but my own family had to come first.

My Mom wanted to keep Paul while we went looking for a place to live. I didn't know why I believed she couldn't take care of him. I went home and explained all the diet to her and how to mix the milk. " As if she didn't know" I knew she would only feed him what was prescribed, so I had no worries there. She was an angel. It was quite a thing for her wanting to do this. That was what my Mother was like, always there in a time of need.

Moving day finally arrived. We had to drive Paul to his grandmothers and get ready for our drive to Toronto. My sister lived in Mississauga so we stayed with her and her husband. The next day we drove to Hamilton to look for an apartment. It was a big city. We had to go back and fourth and few times before we found a suitable place. It was a nice place we found. I was feeling sad as I missed Mom and Paul and wondered how he was doing. We called and everyone was fine, or do she said so. My Mom had a different way of not telling you everything. It would only be until we got settled and then Paul would come back to me. She was concerned about the small amount of food; he was getting, and that he seemed hungry. I knew he was hungry but there was nothing I could do about that. The diet was a strict diet and her had to stay on it.

Our furniture arrived, and we got moved in. within two weeks my Mom came up with Paul. It was great seeing them. I missed my little boy. He seemed to have lost the vacant look in his eyes, and he could smile. The first time since he was born. It was wonderful.

My sister Sharon and her husband Weldie came over about once a month for a visit and we would go to their place. It was nice, as it got me out for some time and to spend with them.

We had our first appointment with Dr. Hanley at the Sick Children's Hospital. He met the our sons and took their history, and they had their blood work done. We were to go there every two weeks for some time. They would call me if their levels were not in the safe range, and their diet would have to be adjusted. This would be scary because they weren't getting much food as it was. I prayed this would not happen. I wanted to know if they could have fruit, other than the jars of baby food, but they couldn't. There was a recipe for bread that they gave me. It wasn't made with yeast. I could give them half a cut of toast for breakfast. This was great news; I couldn't wait to get home and make the bread. What a disaster that was. The bread was so hard you could bounce it on the kitchen floor. I cried and cried. Consequently, so much for toast for breakfast. You cannot make bread without yeast.

I was feeling pretty good. The boys were getting well taken care of by their doctor, and I was feeling confident. Paul had the brain damage done, but it was Kevin who needed all my attention to make sure he would be okay.

Jake had to come to Toronto, as I wouldn't be able to drive myself with the two kids, so time was arranged for him to be off work. After a few months, they gave me a card with three dots on it. This card I would use to dot their blood on and send it to the hospital. This way they could keep a close check on their levels. It was a good idea, but who was going to prick their little fingers, me of coarse. When I would do this once a week, they hollered so much you would have supposed I was cutting their legs off. They never got used to it, nor did I.

Paul had a new doctor who would look after his feet. The bar and cast was off now. That was a blessing. I still had to exercise his feet, and he was put into orthopedic shoes. He was to have surgery soon. The tendons in his legs were tight, and they would have to be done, so he would be walking on his toes.

Kevin was a joy to have. He was a good baby and hardly ever cried. Paul liked him but didn't pay much attention to his younger brother. Paul was not interested in toys, and he didn't know how to play or anything. He was not a very happy little boy, and I felt for him.

He started to walk when he was four year old. His little feet were still turned in, but he managed okay. He was so backward in doing things but that was to be expected. It made me happy when he did it on his time. He seemed to progress some. It was almost like the diet was improving him, and I am sure it was. The phenlalanine levels were good, and I was pleased. I was doing my job well and finally seeing some results.

Paul was still quite a handful. He had quite the temper, and you had to be with him all day long. The head banging went on, and I believed he knew that it upset me. Whenever I stopped him from doing something, he would lay on the floor and bang his head. I never saw Paul without a bruised forehead. He would sit on the chesterfield and rock back and forth for a long time. When I stopped him, he would cry and bang his head on the floor again.

He would climb up on the windowsill in his bedroom. I didn't know how many times he did that before a neighbour saw him and ran over from the other building to tell me. We had to board up his window before he went out of it. Sometimes you wished you had eyes in the back of your head. He would go in my room and slam Kevin's crib up against the wall and Kevin would wake up terrified. I could go on and on forever with the things he did.

He had surgery on one leg, and he would be in a cast for a few months. The surgery went well according to the doctor. When I went in to see him, he was standing up in his crib banging his head on the steel bars. I cried so hard, and I asked him not to do that. What could he understand? I stood by his crib and put my arms around him to hold him and comfort him. I believed I had heard Mama. Was I hearing things or did he know who I was . I never heard it again until he was 10 years old. He would be not uttering a sound.

My sister Sharon was having a birthday party for her daughter Paula, and wanted to know if I could bring the boys over. I was concerned about what he would eat. It was difficult because all the other kids were having cake and ice cream, and he could only eat a few teaspoons of pears and his bottle of milk. He really didn't notice what they were eating. He couldn't comprehend that. I tried many times to wean him off his bottle, but he just wouldn't drink out of a cup. Also, I could not toilet train him. He just would not do it. He was 4 years old and still in diapers.

Paul was banging his cast off the floor or on the walls of his room. Every week a new one would have to be put on because he would break it into pieces. The people who lived below us complained a lot about the banging, and I didn't blame them. I did the best I could to avail the problem, but it wasn't any good.

With the help of my doctor, I could get a home with the Hamilton Housing Authority. Soon we would be moving again. Paul and Kevin would have their own rooms, which was good. The house was nice, and we had a big backyard for them, with lots of room to run around. Jake was playing baseball a lot and doing other things I guess that the boys were my responsibility. The trips to Sick Children's Hospital went on and life continued.

Paul heard the ice cream truck coming down the street and saw all the little kids running to it, so he did too. He couldn't have any of the treats, and he cried. Between Paul and I, we never stopped crying. Life was getting so difficult. I thought of a solution to this problem that just might work. I talked to the ice cream man and explained the situation to him. If I could make Popsicle's out of his milk and add a bit of coloring and gave them to him, he could have one, and he would assume that he was getting what the other kids had. The truck would come on the street behind my house, and I would run to catch him and give him the Popsicles. When he came to the front of our house, he had a Popsicles for Paul and my son grinned from ear to ear.

My first encounter with any type of support was done through my doctor. He knew a group of women that would eventually take Paul for one morning a week to assist me a bit. It wasn't much but anything they could offer would do. Two ladies came to my home to see Paul, and they genuinely thought he was such a handsome little boy. Even so, oh, what a ball of energy. I think they knew I needed the rest. I explained that he could not be given anything to eat while he was there, other than some of his milk. They decided it would be a good idea if they made a little nametag for him to wear with his name on it and a note saying not to feed him. They promised me; he would be fine.

The morning arrived when they came to pick him up. I had him already and was taking him out to their car. He screamed and cried so much and held me so tightly that I thought he would never let go. I wanted to take him back inside the house, but they insisted that he would be okay. They left, and I felt so bad, he probably didn't know where he was going or who those people were. Needless to say I was looking out the door waiting for his return home.

I was kind of looking forward to his morning away, but was always glad to see him when he came back. He had his second surgery on his other leg. It was all the same as before.

The neighbors were good to us. They knew of Paul and when I would be outside with the two of them, they always came over to talk. It was nice to have people to spend some time with. I couldn't have coffee with them in the mornings because I didn't have time; Nevertheless, then again I was never one for doing this anyway.

The doctor's appointments were a big part of our lives. The kids were hungry so much, and we could never eat in front of them. It was a sad thing, and I waited for the day when they would come off the diet.

During one of Paul's appointments in Toronto, it was mentioned to me that it was almost time for him to come off his diet. I could not believe what I was hearing. I had planned for that day the past six years, and it was getting close. What about Kevin I supposed,he still had another three years to go. I decided I could only do this, one day at a time. It would be difficult, but I had handled problems before.

I had a cousin of mine, and his wife Ada, who lived not far from me, and she said that she would look after Kevin while we took Paul to Toronto. I made a big bag of peanut butter and jelly sandwiches to eat. When the appointment was over, and we were driving on the 401 back to Hamilton, I decided to open the bag and give it to him.

He held the bag, and I told him to take one. He didn't understand so I put a sandwich in his hand, but he didn't want it. He really didn't know what to do with it. I was so disappointed, and I couldn't believe what I was seeing. No one told me he would have to be introduced to this different way of eating. I gave him a bottle of our milk, just to try it, but he wanted his own.

The next month more or less went by with no major problems. Paul was being Paul and Kevin was his quiet self.

My headaches were so bad. I was never one to take medication, so I more or less put up with it. The stress was getting to me.

The doctor believed it would be a good idea to have Paul assessed at a hospital in London, so I agreed to have him done. Jake brought him to London, and I said my good byes to him, as I didn't have the heart to go, I believe a friend of his went with him. I didn't know how long he would be gone and considered perhaps a few weeks. They would call to advise us when to pick him up.

The home was quiet with him gone. Kevin and I went out for walks, and it was nice to spend time with him. I was lucky he was such a good child. He would play for hours with his toys. There were always kids around to play ball games with.

One afternoon the little boy next door asked if Kevin could go in house to play. I wasn't sure about that because I never let him stray too far from sight. I asked his mother if it was okay, and she said " yes ." I told her that he couldn't eat anything, but she already knew that. About half an hour went by and Muriel came running over with Kevin in her arms, scared to death. He had red stuff all over his mouth, and it looked like blood for a minute. It wasn't, it was the remnants of a cherry pie. She left it on the table, and he grabbed onto it. Its strange how kids left unsupervised, especially with PKU, can get into such funny but dangerous situation.

I called the hospital, and they said that Kevin would have to go right in to have his blood done. It would have to wait until the next morning, so over we went.

The results showed a slight rise in slight levels but nothing really to worry about. The doctor said he probably had more on his clothes and his hands than he did in his mouth. I believe his friend got more of the pie than Kevin did. I could never trust anyone with him, as it was a dangerous situation.

Three weeks had gone by, and I got a phone call from the hospital in London. The doctors required to see us the following week. The appointment was made and the day came to go. I was afraid, but it had to be done.

When he walked into the office, the doctor passed me a box of Kleenex and said you would need this. His opening words were " One hour with Paul is like 24 with a normal child." " how do you do it?" I was asked. My answer was " who else was going to do it." Paul had a lot of tests done, he was at a border line retarded state. God how I hated that word. It would be worth my while to have him institutionalized as soon as possible. I couldn't believe what I was hearing. I didn't want to send him away. He was my little boy and for almost seven years I looked after him. This was a stupid idea. I got up to leave and wanted to find Paul and take him home and get him away from these people. I didn't want anymore to do with this. Jake requested time to figure this out, and he would call back in about a week. They wanted to keep Paul there, but he was going home with me with or without their consent.

It was troubling for three or four days. Jake said it would be the best thing to do for Paul. Start him early and just maybe, trained people could improve him. He would be in an environment with kids like himself. How by the name of God could he even think this way? This only child knew his mother. This was insane. I could never go on without him. He was my whole world, and I had given everything I was willing to keep on going, but it would not happen my way.

Jake called the London hospital and said that he would have Paul there whenever they said. My heart was broken in two. I couldn't cry anymore and there were no tears left.

The day came when Jake and his friend left for Cedar Springs in Bleinheim, Ontario with my baby. I hardly remember what I did that day, everything was like a blur or maybe even in slow motion. I would never be the same again. Kevin knew his brother wasn't around. He was only 4 years old, but he sensed it. He was talking now and was potty trained. He did it on his own. He was growing up and everything seemed to right on track. I was proud of myself and for all I did for him. I loved Kevin too, but Paul was my first, and I had him a lot longer. I knew in my heart Kevin would turn out normal, and I would just take good care of him, and he would be okay.

My heart was so heavy. I didn't want to get out of bed. I was so lonely and sad, nothing really mattered.

When Jake got home that night, he walked in and asked why I wasn't outside with the neighbors, and after all it was a gathering for us., I wasn't up for any party, and I let him know it.

I wasn't interested in what he had to say about his trip; nothing was of importance tome. Part of my soul had been ripped out.

The days and weeks went by, and I just had to go see Paul. Jake took me up.

In addition, I cannot remember where Kevin went, it didn't seem like he came with us. My memory is lost over a lot of things during that time and my mind was only centered on Paul.

It needed a few hours to drive there. In the distance I could see this big building, and it reminded me of a University. It had lawns around it and lots of trees and flowers. It was a very peaceful setting. We went to the information desk and before long someone had come to bring us to Paul's ward. My heart was pounding so hard, and I was so anxious to see him. There were probably about twenty children on his ward so it took a few minutes to find him. When he saw me, he had the biggest smile. I was sure that he recognized his Mother. We talked to the staff about how he was doing. We took him out for dinner and in town to do some shopping for him. He still wasn't toilet trained, nor could he speak, but he seemed to understand what I was saying.

He loved music and Jake Cash. I would play it for him a lot when he was home. I had a casette tape, and I played it in the car. He looked at me and smiled, recognizing it. He started to hum along with the music.

We only stayed for a few hours, and then left to go home. It was hard leaving him, and I cried a lot on the way home but I knew I would be back soon.

Life was hard for me, and I really didn't know up from down. I had Kevin to take care of but I never felt the bond between us that Paul and I had. The trusted friends I made on my street were kind and caring people, but I was so withdrawn, I just wanted to be alone. I never had a life outside my home really. Sad to say, but that's what I had chose.

Kevin was progressing well and his check ups were coming along good. He was hungry and cried a lot for food. As he got older his food was increased a little more, but it still consisted of strained baby food. There was never a variety for him.

My second visit with Paul came up, and I went by myself. We did the same, pizza for Paul, shopping and a drive. He always liked the car. I spoke to his counselor, and he said that Paul was doing okay. He was having some trouble chewing his food, and they had to cut it up for him. He liked different things. He was beginning to learn his way around the ward. They tried toilet training him, but he couldn't learn how. As the months went by, I was starting to come around. My tears were less and Kevin, and I had a good time together. We would play outside and go for walks. He has some little friends he played with. He was never to be out of my sight though. When he was six he came off his diet, and we were happy. He would still need his blood work done. It was great having him sit at the table with us to eat his meals.

The visits with Paul were monthly, and it was a happy time to see him, but a sad time to say goodbye.

One day Jake came home from work and said that he was posted to Greenwood, nova Scotia. I didn't know what to believe. What about Paul. He couldn't be moved there, and I couldn't leave him on Ontario and move so far away from him, so what was I going to do. I was starting to make my way up the mountain only to find there was a higher one ahead of me.

Jake left, and I was to follow when he found a home to live. A month went by and the dreaded call came. The movers would come and Kevin, and I would go to my Moms by train. I had to see Paul, as this would be the last time for some time. I didn't know really when I would see him again. A friend offered me his car, so I could go and say well to Paul. The tears I cried on my drive to see him just wouldn't stop. I didn't know what kind of shape, I would be in when I arrived there, but the staff knew how I was feeling. Paul knew me, and I ran to him and held him so tight, I never wanted to let him go. I sobbed. My heart out, and he looked at me and put his little hand to my face. I guess to comfort me in his own little way. He was the most precious thing in the world, and I really didn't know how I would ever survive without him. How could a little boy break a Mother's heart? Why wasn't I stronger?

I asked him if he wanted pizza, and he smiled yes. I didn't want to go into town, so we went to a restaurant in Cedar Springs. I guess Paul was there before because everyone spoke to him. We sat down but the tears wouldn't stop. I was embarrassed. The waitress asked me if I wanted a table in the back, and I said that would be a good idea.

The precious time was passing too fast. I wanted to take him and run away somewhere, just him and I, where no one could hurt us again. I loved his smile, his only way to communicate with me.

After we ate, we went back to the hospital and sat on the grass and talked. I told him I had to go away for a while, but I would be back as soon as I could. I hoped he knew what I was saying, and It was doubtful, but I had to say it anyway.

When we left to go to his ward, he took me down hallways I didn't know existed. I assumed for a minute we would never be found, but he knew exactly where he was going. I was amazed; I stayed on the ward for some time until he settled in. Bob, his counselor came over, and we talked. He promised me; he would take good care of him. He gave me his home number and also the number on the ward. I was to call anytime I wanted, and he also gave his address and Paul's as I had planned to write to him.

Bob asked if I wanted to eat supper with Paul before I left, but I said no, that I had to get home. I hugged and kissed him good-by and left without looking back. If I had of turn - I would have taken him home with me.

How could things go so wrong? The misery I was feeling I didn't think I would ever end. Would our lives ever get any better or would we always have a mountain to climb. They jonly got bigger and bigger, and I just kept getting weaker and weaker.

I picked up Kevin at my cousins and came home. I tried not to show my sadness. I talked to Sharon on the phone and told her I would be leaving for Mom's in a few days. She said she was working so I didn't have time really to go over and see her and Weldie. Consequently, much to do with the move, and so little time. She promised she would go up and see Paul. I didn't think she would go that far by herself on her day off. She had Paula to take care of but she did go up and even brought him home for an overnight stay at her home. She was so much like our Mom, and I was lucky to have a sister like her. She said to come to Ontario anytime, and she would take me up to see Paul. I was to eventually take up her on that offer many times. It was all arranged, and I knew she would never let me down. They have busy lives, but they were always there for me. I was lucky to have a family like; I did.

I didn't tell my Mom a lot about my life. God knows she had enough worries. Sharon and I left home, as soon as we were married, but my brother never left Cape Breton. He was lucky to be close to our Mom and Dad.

Before I left Hamilton, I saw an ad in the paper for a little Pekinese puppy. I asked my neighbor if he would drive me out to see it. When I saw this dear little pup, I fell in love with her. I held her in my arms, and I knew she was for me. She cost $75.00 and that was about all I had, but I didn't care. I would make do. She cried all night, missing the other pups I guess. She settled down the next night. Kevin liked her. I named her Tammy after the troubled country singer Tammy Wynette, whom I liked so much.

I had a little cage for her; she would be on the train with us but in the baggage car. When we arrived in Turo, my Mom and Dad and her sister came up to meet us. It was wonderful to see them. It seemed like a long time had passed. My luggage was taken off the baggage wagon, but I couldn't see Tammy; I didn't say anything to Dad, but he noticed I was looking for something. He said he had all the luggage tickets, but I had one in my hand; he didn't know about. Another baggage wagon came when I saw Tammy, I told Dad to get her for me, and that she was mine. My Mom and Dad loved animals, so they liked her. She was so tiny you could hold her on your hands. There wasn't much room in the car, and we looked like the Beverly Hillbillies.

The weekend after I got home Jake came down to see us. It was nice to see him. I had missed him and Kevin did too. I talked to him about my leaving Paul, how terrible it felt and how I missed him.

He said once I got to Greenwood things would improve. I was to be home for a few weeks before our furniture came so I had lots of time to rest and relax. Nobody asked me questions, and I was glad because I didn't want to talk about it. Kevin had his cousins to play with, and I had Tammy. She was very special to me and was with me for 16 years. Maybe she would be some kind of replacement, I didn't know. We moved to Greenwood finally. Kevin would stay with Mom until we got settled, and then she would bring him up. I didn't know what I thought of this place, it was in the country, and I wasn't used to this. We had a small bungalow, and it was nice. I decorated it like a little dollhouse. I loved making my places nice that I lived in . Jake took me over to the base and showed me all around. This would be my home for a few years. I tried to look happy, but I was dying inside. Paul was so far away, and he would never leave my thoughts. Jake did all he could to aid me. I was sorry for being this way, but I just couldn't help it.I had terrible headaches all the time and thought it would be time now to have it looked into. I made an appointment but didn't say anything about my life or what was happening. The doctor said it was always stressful moving to a new place. He had treated a lot of military wives. He gave me a prescription for Valium. I hated taking pills, but I considered I would try it, what was there too loose? I would go back to meet him in a month. The pain in my head was on the left side and felt sharp pains going through my head. After a month I went back, and he gave me more Valium to take.

Kevin came home and was happy. He could adjust to anything. He liked his room that I had fixed up for him. School time was starting, but I found out he would lose a year because his birthday was in October and it had to be before September. I guess that was a rule there. There weren't any kids around our street for him to play with so we did things together. He never gave me a day of trouble; he was such a good boy.

His father was playing ball so I would hopefully take him to his games, and he enjoyed doing that. I would take him home after the game but Jake would go to the Junior Ranks Club with his team-mates. He would firmly stay there for a few hours.

I made friends with Jim and Shirley who lived next door and also Mary and Gerry. They were the only people I bothered with and as the three years went by they would become our good friends. We would go to the dances with them at the Corporals Club. The times that I would go, I didn't really enjoy myself as I always had a sad depressed feeling.

After six months of living there I specially wanted to go see Paul. Jake made the arrangements, and I would fly out of Shearwater, Halifax on the military train to Trenton, Ontario where I would then have to rent a car and drive to Chatham where Paul was. It was a lot of planning, but I would do it and I didn't know how I would ever drive across Toronto. I must have had more courage than I do now.

As I boarded the plane, I was scared to death. I only flew twice and that was when I flew to Comox and back home. I didn't like it at all. I had to do it because three was no other way to get there to see my son and nothing was going to stop me. I found two military fellows and asked if I could be seated in the middle seat between them. They were both said yes-that; they didn't mind a good looking woman sitting by them. When the plane taxied out to the tarmac and got ready to speed down the runway, I grabbed them by their hands, and I don't think I let go until we arrived in Trenton. This was terrible, and I hated it. My head hurt, and I had such pressure in my ears, that ii would last for days after. It was a long drive from Trenton to Chatham. I would guess about six hours driving by myself.

I got to the hospital and made my way to the ward. My legs would hardly hold me up. I saw my son sitting at the table eating his supper. He looked up at me and smiled the biggest smile that I had ever seen. I went over to him and hugged him so tight. I missed him more than I believed was possible. I let him finish his supper while I talked to the staff. They said he was doing fine and was always happy when he got my letters, and that they would read to him. Paul took me to his room and showed me his bed. He still couldn't talk, but could communicated in our own way. I was taking him to stay overnight with me at the motel in Cedar Springs. I hadn't made the reservation yet but there was never a problem to get a room.

Paul found our room quite small in comparison to his ward. I explained that he would be staying with me for the night, and then we would go back the next day. I didn't know what he thought, and I didn't want to scare him, It was late that night when a knock came at my door. It was the owner of the motel, and he brought me a tea and sandwich. He knew I had a long trip and figured I hadn't had time to eat. He was so right, but I was too excited before then to think about eating. Paul went to sleep early, but I just sat there for the longest time looking him over. His hair had turned brown and his eyes were green. He had my characteristics. He was such a handsome boy, and I was oh so happy. He would forever change my life without even knowing it.

I went through all the emotions of grief. We had breakfast that morning before I took him back to the hospital. The visit was all to be short, but they weren't all like this. Most visits were only for an afternoon. I said good-by and left. When I got off the plane, the ambulance was waiting for me. I had terrible pains in my ears and head and during the flight the pilot had to descend to a lower altitude to ease the pressure in my ears. For a few days I was in a lot of pain, but it would eventually leave. I was worried about flying and having to go through this every time I got on a plane.

I had a terrible time when I came home, I didn't think I could take this any longer. The continuous depression was getting worse, and I just wanted to die. I guess I wanted the pain and the sadness to go away. There was such a void in my life.

The days ran into weeks and the weeks ran into months. Jake was curling a lot, and he became the manager of the Corporal's Club on his time off, which kept him out a lot. I couldn't blame him for wanting to be away.

I was concerned about Kevin and not having a father around. He didn't do anything with him. He made friends when he started school, so he had kids to play with, and he would always have a Mom who loved him. We were becoming close and when I tucked him in at night, I always made him say a prayer for his brother. He was always interested in my visits to Paul, and we would talk about them.

I started taking more Valium, and I felt like I was floating through the days. Every day would be the same. I didn't want to talk to anyone, not even my friends. Mary dropped over and asked if I could look after her little boy until she could find a sitter. One look at me and she had her answer. She said she would be over later.

That day I consumed a lot of pills. I was out of control. She did come over later that night, but I didn't even know she was there. She took Kevin to her place and took me to the hospital. They pumped my stomach and put me on suicide watch.

The next day I didn't want to see anyone. Jake came, but I couldn't even look at him. I was disappointed in myself and wondered what Kevin was believing. How could I be such a bad mother when he was deserving so much more?

The depression would be around for a long time. I went to my parent's home a lot on weekends. Kevin would come with me and was always happy to see them, and I was glad to get away.

I had an appointment at the hospital for my head aches. They did x- rays and found that I had a TMJ problem. (Temporomadibular Joint Syndrome). It was more of a dental problem rather than a medical problem. I would have to go and see an orthodontist. An appointment was made, but I had to wait 3 months. The day arrived and off I went to Halifax. The dentists examined me and noticed I had a lot of clicking in my jaws. I was always grinding my teeth and unconsciously grinding them. It was worse under extreme stress. I was told that whiplash injury, which had resulted in this condition. We went to Hamilton.

Jake was looking for an apartment when a car struck us while we were stopped at a red light. The whiplash injury was very painful, and I had to have therapy on my neck and had to ear a neck brace for a long time. It ended up in a law suit.

The orthodontist made personal impressions of my teeth, and I would have to go back in a few weeks to get my splint. I would have to wear this brace all the time especially at night. The splint would aid to elevate the pressure on my jaw from biting and grinding. I would eventually end up wearing it for 40 years. The splint was expensive, and I would require at least one or two by the year. It was difficult to talk or eat. It was a bite plate made of plastic, and it would fit over my bottom teeth. I eventually got used to it. It gradually helped reduce the pain in my head, but they were no cure. I had to have check ups every so many months.

My visits with Paul continued. I would go about twice a year. He was growing up and could say a few words, one being Mama. I was at peace. His counselor would take him on camping trips with his family and always had pictures for me. I always felt so much for him, he had my trust, And Paul was very lovable, and you couldn't help but like him. He was so good looking and normal looking. You would never know he was mentally handicapped until he tried to speak.

After a few years a posting came for us., We were going to Sydney, Nova Scotia. Home at last. This couldn't have come at a better time, and I couldn't wait to go home. This was the first good news in years, and I knew then that someone was looking out for me. Mom and Dad were so happy.

We bought a new mobile home, and I was excited. I would be moved to the base in Sydney where we would live for the next four years.

It was hard saying good- by to our friends, but this is what happens when you are in the military. I would always be in contact with them over the years.

Shirley and Mary both passed away in the early 90's from cancer. I was far away from Greenwood, then so I couldn't go back. I wrote to the families and expressed my sympathy. I would never forget them and all they tried to do for me. I was filled with sadness.

The day came for us to leave. It was the first posting I was ever on that I was excited. My Mom and Dad and brother were there, and it was all I wanted. I wouldn't have an adjustment to make this time. It was a small radar base with only 130 personnel.

Our mobile home arrived, and I was busy setting everything up. Mom came and assisted. She was very happy and knew things would be better because they were there. She hopefully took Kevin on the weekends and spoiled him. He always looked forward to Friday's.

. He was a wonderful son, and I was lucky to have him. He was growing up fast and his progress was normal. While he was on his diet, I made sure he wouldn't get anything to eat that he wasn't supposed to have. Paul could have been like him had he been diagnosed when I noticed he was different.

Six months had gone by, and it was time for my visit to Paul. Jake had strongly thought by now I would have forgotten all about going, but I didn't. Paul was still very much on my mind.

I would travel by train to Toronto and Weldie would pick me up at the station then Sharon, and I would drive up to Chatham. We would spend the afternoon with Paul and then leave for home. It was always so hard leaving but not like it was.

Paul was doing fine and progressing as fast as he was able too. It was a long trip from the time I left Sydney and got back home again. Mom was always there to help with Kevin.

After I came home, I applied for a job at Woolco Department Stores. I was hired and got a job in the furniture department. Everyday I would go to my Mom's for dinner as she lived nearby.

Jake was gone a lot, but it didn't matter. I had my job and when it came home, there were a lot of things for me to do. Kevin was doing well in school and never had to be told to do his homework, it was always done. Him and I were getting very close, and it was a good feeling I had.

Jake's Mother and Father lived in Sydney, but I didn't go there a lot. I had a busy life now with working and everything. There were parties at the mess, and we went to a lot of nice functions. Id didn't get close to anyone at the base. Time went by, and I eventually took two trips a year to see Paul and always wrote letters to him.

Several years had gone by when we were posted to Calgary. I didn't want to go but pretty much had to.I would be so far away from everyone .Calgary was a nice city. I loved going to driving to Lake Louse. The mountains were beautiful.I would learn to ski there and enjoyed it. After a year Jake would go to Egypt for a six- month tour. He called every day or two on the ham radio. That was an experience, but at least I could talk to him.

When he came back from Egypt, I decided to go to Toronto. Mom and Dad were there when I arrived at Sharon's. They wanted to see Paul, so they supposed this would be the time to go. We all drove up to Cedar Springs. It was a long time sine they had seen Paul, and they couldn't believe how big he was. My Dad cried a lot. He was like that, and I would be like him. We had a nice visit. Paul was happy to see me. He was so far away now so I could only go once a year to visit, but I was always in contact with him.

After a few years in Calgary, we got posted to Kingston, Ontario. I was elated. At least I would be in the same province as Paul and that was a good feeling. I didn't have any regrets leaving Calgary. I really didn't care much for the city. When our furniture left, we drove to Cobourg, Ontario where Sharon and Weldie were. They went camping there every weekend so when we arrived, they were there. We had a camper trailer set up by their cottage. I met a lot of their friends, people who I would occasionally meet over the next 20 years. Most of them were from the states, and I loved their accents.

We briefly stayed there for three weeks' when Jake got a call that our furniture arrived so we left the next day for Kingston.

I liked the house; it was three bedrooms and lots of room. I was happy for the first time in a long time. I knew I would be happy here, it was a feeling I had. I couldn't wait to go see Paul, I had to get settled first, I didn't want to start any trouble.

Kevin was fine, and I willingly helped him with room. I had bought him all new furniture, so he was pretty excited. He liked his desk and would do his homework by himself, so I considered it was a good investment. I was proud of him and how well he gradually progressed over the years.

Jake had joined the curling club. That was to be a big part of his life. He fished and golfed all summer. He bartended at the curling club. He was never around too much, but always on time for his meals. Things were pretty tense. It seems that he was all for himself, no one else mattered but him.

During the hunting season, he went to Sydney for a few weeks to hunt with a friend of his. This he would do every year. We went camping the odd weekend. I got a job at the Shell gas station on the base. It was weekend work, but I didn't care. i needed to get out of the tiny house. Kevin didn't have any friends, they were all younger, and I felt sorry for him. His father could have been a big factor in his life but never was.

I left for Cedar Springs myself. It was a long drive, but I did okay. It was nice to see Paul, he was 18 years old now. He could say a few more words. He loved his music and would listen and hum along with all the songs.

Sharon and Weldie had gone to Mom's for Christmas. On their way back New Years Eve, they were coming in for supper. I cooked a turkey and had everything ready when they arrived. Things were rather quiet that night I noticed.

They left for Toronto around 9:00, and I told Jake that I would get ready because we were going to the dance at the curling club. He said no, were not going, and that he wanted a separation. He told me that Kevin, and I could remain in the tiny house until May, and then we would have to move. I couldn't believe what I was hearing. I wasn't always the perfect angel, but I never thought it would come to this. I cried all the time.

He moved his clothes down to the room downstairs and the only time he came home was to eat and change, and then he was gone again. I did his clothes, his uniforms, and always had supper ready, I assumed this would pass. I started taking pills and ever wanted to get out of bed. I would always do my house work because I loved a nice clean place. I would drag myself out of bed just to get my work done. Then I would pop pills for the rest of the day. I still did my job at the gas station on weekends.

It was a lonely time for me, and I was so depressed. I was scared because I didn't know what would happen to Kevin and myself. I wasn't making much money. I worked on Saturday and Sunday at $2.90 an hour. I always took care of my family; therefore, I wasn't trained for anything that would make money for us to live on. I was worried.

Jake would come home after he knew I would be in bed. Everyday I would get up and find Anne Murray records all over the place. Kevin and I hated Ann Murray. Jake talked to Kevin for one minute, no longer ad told him he would be leaving in May. He never asked him to go live with him or what was going to happen to us. He didn't give a damn about his youngest son. He never did, so why start now.

I talked to Kevin and told him things would work out somehow, they always did. I said it, but I was scared. I just didn't know where to start.

My depression was so bad I didn't want to live. I was taking pills and living in a haze. A neighbor across the street must have heard something because she came over to see if she could graciously help. She realized things were bad, and she believed it would be a good idea to take me to the hospital to get some assistance. I must have gone along with it because I would remain there for many months. I wouldn't eat anything for days and was told if I didn't they would feed me by needle. I tried to eat something just to keep them off my back. I was loosing a lot of weight. I wouldn't get out of bed and go to the dining room where patients are. I wanted to be alone and wouldn't talk to anyone. My psychiatrist came once everyday and would do a lot of talking. When I heard enough, I just got up and went to bed.

People were turned away at the desk because I didn't want to see anyone they had to honor my wishes. I must have been a difficult patient. I guess I couldn't help it. I just wanted to stay In the hospital and never leave. I felt protected there. It was hard living in a house where there was no talking.

Time went by and once in a while I would leave the ward and walk down to the store for cigarettes. It was scary out there, and I walked with my head down so I wouldn't have to see anyone. Kevin would come in at night and see me. He had given me a letter that Jake's Mother had given him. He had no intention of taking Kevin with him.

Kevin asked me to come home, and that it wasn't the same with me gone. I promised him that I would and kept my word. When my doctor came in, I told him I had to go home and that my son needed me. He said the only way I could go home was if Jake left. I asked the doctor is he would call Jake, and he said he would. When he came, he said that he would move out as soon as he found a suitable place. He told Kevin he would be moving out the following Saturday. That Saturday morning I was released. It was strange being home. All the pictures and ornaments that Jake had sent me from Egypt were gone. Pieces of furniture like the middle unit to my wall unit. The end tables to go with my coffee table and a lot of different things. As he was taking things out of the home and loading up a truck, he came in and kissed me and told me that he loved me. I could not figure out for the life of me what just happened.

Kevin and I held each other and cried. I knew he was hurting, and I would have to be strong for him. He looked so alone. Jake might not have been a good Dad, but he was still his father.

I was offered a full time job at the service station, so I decided to take it. The pay wasn't good, but it would assist. I found an apartment for Kevin and myself. Sharon came down to willingly help me unpack. It seemed like I had million boxes, and she laughingly said she would never help me move again.

It was different moving into an apartment after living on the base for all those years. Tammy, my little Pekinese died on Christmas Day. What a hard time that was. I still had my collie dog, and he would have to move with us. I didn't know how he would like living in an apartment, but I couldn't give him away to anyone.

Jake sold our camper trailer and gave me the money to buy a used clunker. Wasn't living near where I worked, so I needed a car. He would pay me just enough money to pay my rent. It was hard on me not having much. My lawyer I had through legal aid wasn't much help. She didn't tell me I could have gone after his military pension and his Canada Pension. I must admit though he did pay for the divorce.

In 1987, my divorce went through. We had been married 20 years. There was a big vacant slot in my life, and I wondered if I would ever be happy again. That year I briefly went home to visit my Mom and Dad on the train. Kevin was camping with Sharon, so I didn't have to worry about him. I was home for about a week when on m way back, I met David on the train. We talked a long time and had breakfast in Montreal together he was a wonderful person so caring and attentive to me. I was afraid to ask him if he was married. I wanted to know, but I didn't want to know. I was getting off in Kingston, so he asked me if I wanted him to stay the weekend and go to Toronto on Sunday where he lived. I believed that this was a good plan, so he stayed with me that weekend and just about every weekend after that. We talked on the phone all week and wrote letters. Kevin met David, and they got along well.

In December we became engaged. I was in no hurry to get married at that time. My divorce ended four years before I met him, but I was in no rush. He liked Kingston and was considering about moving down and finding a job. I knew then I wanted to spend the remainder of my life with him.

One day I got a phone call from Cedar Springs. They wanted to know what I would strongly think about having Paul moved to Piction, Ontario which was no more then an hour from me. Apparently, there was a client in Picton, whose' parents wanted her to go to Cedar Springs in Chatham, so they would just have to do a switch. It was wonderful to have Paul close to me. It was so unbelievable, and I was so happy. I wanted him right away.

Twenty-five years of always traveling to see him could be coming to an end. This was what I wanted all his life. I didn't want to think about something going wrong, it just couldn't.

A few weeks had passed, and I got another call. Paul would be down in two weeks for a trial visit. I would be there for sure that day . David made sure that he would be there with me for this special day. For two weeks I couldn't eat or sleep. I prayed so hard that nothing would happen. I would have been devastated.

The day arrived, and away we went. It was an old military base where clients stayed in the hospital. We were taking to his ward when I saw him, he looked so frightened. I just took it easy with him. I guess he didn't know where he was and all these strange people around. I held him in my arms and felt for the first time in years, that he would never leave me again. All the emotions I had just cam pouring out. It was such a relief. David and I talked to Paul and explained to him that he would have to go back to Cedar Springs to get his clothes in a few days, and he would be back.

My life would be different now for the first time in a long time. How this little boy could control my life and my feelings was something else.

Kevin decided he was going to join the military in 1985. We talked about it some many times. I didn't know if he would be able to get though basic training. He was such a shy timid person. He took his standardized tests at the recruiting center and passed. I was glad for him. When his leaving date came to go to Cornwallist in Nova Scotia, I felt so sad that he was leaving home. He was a young man now, and it was time to go on his own. No Mother wants her child to leave home. He was the only one who held me together all those years.

David moved down to Kingston, and it took away some of the emptiness I was feeling. Kevin wrote to me often. He was finding it hard, and I knew that he would. David gave me a small recorder so that I could tape our voices on it, and It would give him a bit of moral support. Every night I got home from work; we would record for him. He was always happy to get the tapes and said he played them again and again. I knew we were onto something. This would continue all through his training.

My Mom and I went to his graduation. I would go by train to Sydney, Nova Scotia to meet her. We borrowed my brother, Rodgers car and drove to Cornwallist. Kevin was allowed to leave base for that night, so he stayed in a motel with us. It was wonderful to see him; he changed so much. I drove him back to the base early the next morning, so he could get ready for his graduation. My stomach was doing flips but my heart was bursting with pride.

We arrived at the parade square and took our seats, there were so many recruits that I never knew how we would find him. My Mom said " why doesn't he wave," I just looked at her and smiled . I hadn't seen him in uniform before and all the guys looked alike. Kevin was 6.1" so I just looked for the tall guy and spotted him in the last row. The marching display was great, and I realized then how much training they would have to learn to pull something like this together.

I was so proud of him that the tears flowed a lot that day. Not tears of sadness this time, but tears of joy. There wasn't too many of those in my life.

I was anxious to get back to Kingston and await Paul's moving to Piction. I was having terrible headaches all the time and was under the care of the psychiatrist, who I had when I was in the hospital. I would go to him for the next 15 years. He put me on a drug that would help me sleep and ease the pain in my head. I was diagnosed with chronic head pain. I had acupuncture on a weekly basis, which didn't help much.

I had used a tens unit, which didn't help. I was sent to many neurologists fro examinations. It would end up that I would just have to live with the pain.

It was difficult to get my head off the pillow in the mornings to get ready for work. Many days I had to drag myself just to do just that.

David was working at the Royal Military College and things were easier for me financially. Jake didn't leave me with much when he left. He was the one who ended up with the big house, the swimming pool, the cars and the boat.

Kevin came home on his first leave. He had finished his training and got a posting to Calgary. He was happy because we had been there before. I cooked all of his favorite food while he was home, and he enjoyed that. David and Kevin got along fine, and he would always be there for Kevin. He was a typical young man; clothes all over the apartment, always in the fridge eating, bringing home Harvey's burgers one hour before supper. The television was always on sports while we hopefully found something else to do.

His running our place would continue everytime he came home, but we didn't say anything. He was on leave and we wanted him to enjoy himself.

The time came for him to leave. I cried a lot when he left, but I knew he would always come home to us.

It wasn't long after this that Paul moved. We would go to Picton the following weekend. He settled in.; He always knew who I was. I never gave him time to forget me.

We would take Dusty, our collie dog, up and Paul would walk him all over the grounds. We would have picnics and barbecues with him. The town was small, so we had to make our own fun. We would visit him just about every weekend over the next decade. The winters were rough, but we were always there. He came home on all the long weekends and holidays. It was nice having him home.

David always kept him busy. Paul wasn't one for playing with anything or watching television; he had no interest. David would take him out for drives and always tried to keep him busy. He was wonderful with both my boys. All the years that Paul was in Picton, maybe twice a year his father would go see him. He only lived an hour away, but never bothered.

When David would bring him home for Christmas, Jake would call and ask if he could take him out to his place for a few hours. He never offered to drive Paul back to Picton, David did it all and never complained once. David has three children of his own that he has never seen in years because their Mother said that their father had died. One day I hope that they will come looking for him.

He is a terrific Dad and my boys love him dearly.

In the late 80's I had to quit my job due to the headaches. It was a job I had for almost ten years, and I loved it. I worked alone and it was hard to get someone to fill in when I couldn't make it. I went on lifetime disability by reason of head pain.

Kevin came home on leave during the summer and always for Christmas. It was only a small apartment we had and when I look at it now, I wonder how we ever did it. He would come home for a month at a time, and he would have to sleep in the living room on a mattress on the floor.

Kevin got posted to London and after that he went to Windsor. David and I visited him a lot on the weekends, and he came home. I was happy that he was closer to us. We would willingly help him put up curtains and lay carpets for him. I generously gave him a lot of things to help him out. He would call me two or three times a week to ask how to cook this or that. He was to become a good. He liked polish food, so I would show him how to make cabbage rolls and perogies when he came home.

He wanted to wall paper his bathroom. David is okay, but I wouldn't hire him out. We found some wallpaper "not flowers" and would put it on for him the in the future we went up. David didn't want any help. He said I was only in the way. After he was done, I went in to inspect it. I noticed that he had one sheet upside down. I didn't have the heart to tell him as it took something like 8 hours to do a small bathroom, so I just let it go.

Kevin left the following year for Edmonton. While he was there he went to Bosnia for 6 months. After 3 months, he was allowed leave to come home for two weeks. It was wonderful to see him, but we saw an obvious change in him. He was quiet and always had to be going somewhere. While he was gone, we sent him care packages every week. He got so much that he would give some to the other guy.

When he came home for his visit, we would go to the airport to pick him up. We made a big banner saying " Welcome Home Kevin" and put it on the window of the terminal. He thought that it was neat and so did everyone else. He would be spending Christmas over in Bosnia, so we gave him some decorations for a tree they had, just to keep the sadness down a bit.

We had to go to a meeting in Picton concerning Paul. I was afraid and didn't know what was happening. They obtained word from the Canadian government that Picton Heights would be closing in 6 months. This meant another move for Paul, and I didn't know where.

Paul loved it there. He moved from the hospital into group homes. He loved the staff, and they were so good to him. He could bowl, swim, and play badminton, tennis and floor hockey. There was a big gym there, and he would go with the staff. They loved taking him there because it also gave them something to do. Paul would talk about the staff long after he left there.

The arrangements were made to have Paul live in Kingston in a group home with the Ongwanada Hospital. I was happy as you can all imagine. We wouldn't

have to travel anymore and we could have him home whenever we wanted. This was too good to be true.
They took Paul down for an overnight visit to see how he would make out. I knew he would miss his friends and the staff, and I felt sorry for him. He was happy to see us when we went down to see the renovated house where he would be living. His counselor came down with him to make the necessary adjustment easier for him. We explained to Paul that he would be living here now, and we could see him all the time and take him home with us. He seemed sad and confused but David and I would be there for him, all the way. There were six other clients at this house. It was nice, big, with a fenced in backyard when they could barbecue. The staff decorated it nice, and it was homey. I knew in time, Paul would fit in. When Paul was ready to move to Kingston, they called his father and told him where he would be living. He wasn't very pleased about him coming to a big city, as if they were going to let him play in the street. During the time he lived in Kingston, his father still only saw him no more than twice a year and only living 10 minutes away.

The staff was wonderful to Paul, and he liked it there. He would go to Crescent Center every day. This center was like a workshop. We would eventually take him to our house two or three times a week for supper. He enjoyed that and David would take him to the mall shopping, and we did a lot of walking. I enjoyed my life now; everything seemed to fall into place. We took him back to Picton for visits, the houses were all closed now. He wondered where all his friends had gone. The clients were his friends for over 10 years.

He was an altar boy at their church and sang along with the priest when he played his guitar. He must have had some feelings after living there for so long. We would always be there for Paul, and he knew that.

Two years had past; it was June of 2000, when I received a letter from Ongwanada telling me that in three weeks Paul's group home would be closing. I couldn't believe whatI just read. I called Paul's home to get some answers, but I was told, " we cannot talk about it" or we will lose our jobs. One staff member knew how upset I would be and did call me. They found out the same day I did. I asked where the kids would be going and I was told they would be moving to Home Share homes. I had no idea what this was. I was told that they were homes the Ongwanada buys, and then puts ads in the local newspaper for people to move in and run these homes. I couldn't believe what was happening. I cried for a week. How could they do this to all these kids and lead us to believe that this would always be Paul's home . I wasn't feeling to well at the time so David had to fight this without me. Everything I dreamed about was coming apart. Every time I felt happy, something bad was around the corner. How could this be? I didn't do something so terrible in my life that I would always be punished. Maybe it was because I loved my son so much. This wasn't easy for him either.

After David did a full investigation into where Paul would be going, he found out that this would not be appropriate. Paul needed consistent supervision and only university trained people would qualify to take care of him. The same qualified people who would take care of him since 1969. As far as we were concerned, this is not what we wanted for our son. We had many meetings but nothing would change. We didn't know what to do and our hands were tied. We had to decide to let Paul go into Home Share; we had no choice. I told them it would only be on trial basis.

Paul always liked consistency in his life. The group homes had regular bed-time, meals at a certain time, baths, everything was regimented. When he went to Home Share, he was out late at nights, and there were times when I couldn't have him home for supper because they were going to a church function. I wasn't very pleased.

The following was a letter written by one of Paul's staff, Linda Willoughby that was sent to me.

SAYING
GOOD-BY

" As everyone knows, Wilson Street Community residence closed this June. It is with heavy heart that we write this note. We are of mixed feelings, sad, apprehensive and concerned, for our family, is the first feelings that come to us as well as why us. After a week of thinking and stewing, we decided to look at this move in a different light- not with sadness but hope. We looked back at our beginnings. In June of 1990, when we started, our first objective was to get used to living in a home, which was quite different from Penrose. The first few years we were busy learning new skills and having many new adventures. We , as staff watched in awe as our clients blossomed and became close to each other, looking out for each other and helping one another in daily living.

We also looked back to many trips and adventures we had. Trip's to Canada's Wonderland, van rides to see houses decorated for Christmas and Halloween, trips to the country, flea markets, and garage sales and of course picnics in the park. On a hot summer's day you may even catch us slipping out after supper for a dairy queen which was not on our diet, but it tasted so good, and then everyone singing to the radio on the way home. These are just a few of our memories that have become special to us.

So we all move into different places. We will miss each other, Marjorie who is the mother hen of the house keeps everyone in line. Heather, for her very cool wit and crazy sense of humor. Colin who loves to sing even though he is tone deaf and cant carry a tune. Jake for his bright smile, wonderful laughs and is the official welcoming committee. Paul, our newest family member who is always willing to help everyone. Katrina who always has her pictures rady to show everyone all of our adventures. Pattie, who likes to keep staff jumping when it is too quiet by coming into the kitchen and pretending to try to take come food not on her special diet and when shooed, giggles like she's playing a joke. Everyone will remember the staff that has, at times, been tucked into bed by clients after a long hockey game when he fell alsleep but they didn't. our orgainizer who keeps everyone guessing where things are, but without it, our home would not be so nice to work in. the old mother hen who will drive new staff crazy by telling them how to do everything even after they have been there for more then 3 months. Our crazy CSW's who have brought so much into our home and have been great to work with. Lastly, our supervisor who has been great and takes all our antics gripes and moans in her stride.

Although we will all go our separate ways, it is not good-bye but so long for now as time goes by we will meet again and we as staff, will take pride in all your new accomplishments. Wilson Street will always be remembered as the place where we stopped at for a short time on our journey in life, where we learned skills, and tons of fun and bonded in a special way."

These people were feeling what I was. It was no wonder I didn't want the home to close.

Paul lived in Home Share just long enough for us to find a two- bedroom apartment in our building. In one month he would come and live with us. What a rude awakening that would be, but we knew this would be the best place for Paul. We didn't have an other choice at the time.

Our closets were a little cramped but we made do. Paul was happy and that was all the mattered. I was a Mom again and believe me it wasn't easy.

A week after he came to live with us, I was diagnosed with breast cancer. This was a big mountain to climb. I was scared to death but after the shock, I was able to start to fight back. I had a lot of hard knocks in my life and this wasn't going to end it for me. My family would be there for me, my brother would call me from Nova Scotia just about everyday to keep my spirits up and Sharon would too. They were the best. I cried a lot and asked again why me? I would ask that question a lot through the years. I guess God only gave problems to those people who could handle them.

I talked to David about Paul, and what we were going to do with him now. I was sick and didn't know how we could manage taking care of him. I really didn't know what was ahead of me. David said we would be okay and that Paul was to stay. I was so relieved.

Kevin was worried too, but I knew that he would be okay. Paul was 38 years old and had the mind of a 5 year old. There wasn't much he couldn't do without help. My doctor said I was a grandmother who had a five year old son. I didn't realize it then, but it was true.

I would talk to Kevin a lot. I wanted so much for him to come home and put his strong arms around me. I needed that. I cried when I talked to him, and it was so hard not to. I didn't want to die. I would miss him too much. He was so special to me.

It would be 31 years before I got to tuck Paul in at night. I always hoped that someone would do that. He would put his arms around my neck and just hold on and say goodnight Mommie. I would say " Good Night- Sleep with the Angels, Mommie loves you". One night I forgot to say it and as I was leaving his room he said " Mom" and I said " Yes Paul'. He said " Sleep with the Angels". He had a way to tug at my heart strings. I looked at him so many times and thought what would have been. He was so considerate, loving and caring. I cannot go there, it's just to painful for me. All I know is that he is special to me. Some days he would ask me if I were going to die and I would tell him, " no way that Mommie would be there for him always". He understood that I was sick and sometimes you would see tears in his eyes. What was that little 5 year old thinking? I tried so hard not to cry in front of him.

In the morning when I would get up and help David get him ready for the workshop, he would come to me and hug me saying he loved me. He never missed a morning and when he would come home in the afternoon, he would hug me again and say, " I missed you." I was so proud of him and my heart feels like it could burst from the love I have for him. All the "only ifs" cannot change a thing.

Kevin met a woman, and soon they moved in together. I was happy for him because he never really had gotten involved with anyone and always seemed to be alone. As time went on things sIt would be 31 years before I got to tuck Paul in at night. I always hoped that someone would do that. He would put his arms around my neck and just hold on and say goodnight Mommie. I would say " Good Night- Sleep with the Angels, Mommie loves you". One night I forgot to say it, and as I was leaving his room, he said " Mom" and I said " Yes Paul'. He said " Sleep with the Angels". He had a way to tug at my heart strings. I looked at him so many times and supposed what would have been. He was so considerate, loving and caring. I cannot go there, it's just to be painful for me. All I know is that he is special to me. Some days he would ask me if I was going to die, and I would tell him, " no way that Mommie would be there for him always". He understood that I was sick, and sometimes you would occasionally see tears in his eyes. What was that little 5 year old thinking? I tried so hard not to cry in front of him. started to fall apart between Kevin and myself. The calls were becoming less and less. She didn't like the relationship that we had and slowly started to driving the wedge between us and I could feel it.

One day a letter arrived, and it was from her. She told me to stop bothering Kevin about my illness. She told me to go to the cancer clinic and see people who were really sick, and that I would soon forget about my cancer. I was shocked and couldn't believe what I was reading I phoned Kevin when I knew he would be home from work. I cried so hard when I heard his voice that he couldn't understand what I was saying.

I asked him if he knew about the letter that she wrote me. I couldn't even say her name. He said that he had read it, and he was the one who mailed it. I asked him how he could let her hurt me like that. All he said was " she wrote it, not me". She said that they were getting married next year, and she didn't give a damn whether we went or not. She didn't have to worry; I wouldn't be going. Kevin was different person than we knew. She lied to him about many things before they were married. I didn't interfere, all he ever said to me was we have to learn through our mistakes. A man 37 years old and a woman of 44 years should have been learning something along the way. I didn't care anymore; they could keep on doing what they were doing. I didn't have any fight left in me as where they were concerned.

It was hard on my nerves waiting 4 months for my surgery. I just wanted this thing out of my body. We didn't make plans for Christmas that year, as the excitement just wasn't there. I was to have my surgery December 19,2000.

That morning I had to be in the hospital by 5 o ' clock and I would have my surgery by 9 o 'clock. The doctor knew what my wished were so I left everything in his capable hands. After it was over I had a feeling that something was gone out of my body. My life had changed a lot since then and I tend to look at things differently now. Things that seemed important are no longer.

When I came home the next day, the VON came in to see me. They certainly are wonderful nurses. They tried to answer all the questions I had.

My arm was real sore due to all the lymph nodes that were taken out. Over the next two years, I didn't gain back much use of the arm. It was very painful to lift it so I tried not to, even though I had to do arm exercises. If all that I am left with is a sore arm, then I can live with that.

I did my best to assist David with Paul, and sometimes I felt that I was in the way. As time went on, things got a little better. Paul was doing fine and David was tired as he had a lot to contend with.

I went through my radiation okay. I had to go five days a week for some time. It kind of wipes you out. The staff in the radiation department is so dedicated; they do everything possible to graciously help you cope with this. My doctor, Dr.Shelly was terrific. She seemed to take my fears away.

Sharon and Weldie had retired and had a beautiful home built in Cobourg which is not that far from Kingston. I was happy because we would see them more often now, and we needed them both. It was so peaceful there, so much different from living in an apartment building. It's a country setting and that's what I like.

Paul had lived with us for over 2 years, and we were getting tired now. He had respite care every weekend with a young couple. They have four boys or their own and included Paul in their family. Paul loved going there, so I knew they took real good care of him, and they were only 5 minutes away from us. Nothing but the best is what we wanted from them. The staff at Ongwanada, Jeff Gifford, Ellen Peters and Charmaine Wood were always there for us. Charmanie being Paul's worker would offer to take Paul to his appointments if I couldn't, and would do so much to aid with Paul, as required. Ongwanda provided respite care when we didn't think we could get through another day. They were all just a phone call away, and we will be forever grateful for their help and support.

Paul is a wonderful son, always trying to please. We taught him a lot since he came to us. He is very polite, always happy and loving. He doesn't ask for much. He loves David, and they do everything together. I honestly don't know where he gets the patience. He treats Paul with love and respect. I am so glad to have found him. They look good together.

Paul will be leaving us the end of January 2003 and will live with Amy. He needs to be around young people now. This time I picked where he would be living. He is 40 years old but very much like a little boy. It will be difficult to let him go and no doubt a very dad day for us. Dave, and I need some time for ourselves now, just to relax and so things together. We will always be there for him, and he knows that. He will probably have more meals with us than at his own place.

The last time I saw my brother in law, he told me if anything ever happened to use that he and Sharon would always be there for Paul, and I know they would. My family, being my brother and my sister were always there for me. I am truly grateful. I always felt I walked through life alone.

My journey has been a long one. I don't think I would want to do it again. I don't know what lies ahead for us, but I am going to try my best to make my life happy and that David and I enjoy the remaining years we have together.

We will make sure Paul gets the best care possible, and that he knows were only a phone call away. He will never know how much he gave to me when I was sick. The hugs and the, " I Love you Mommies" would get me through some very dark days.

What I gave him all his years was nothing in comparison to what he gave me. He is a precious gem.

Good night - Sleep with the Angels. Mommie loves you.

www.ingramcontent.com/pod-product-compliance
Ingram Content Group UK Ltd.
Pitfield, Milton Keynes, MK11 3LW, UK
UKHW041901190726
13854UKWH00003B/1029